# ATLAS OF THE HUMAN SKELETON, 3/E

Updated to accompany

PRINCIPLES OF ANATOMY AND PHYSIOLOGY, 10/E

## GERARD J. TORTORA

*Bergen Community College*

Photographs prepared by

## MARK NIELSEN

*University of Utah*

JOHN WILEY & SONS, INC.

To order books or for customer service please, call 1(800)-CALL-WILEY (225-5945).

ISBN-0-22377-8

Printed in the United States of America

10 9 8 7 6 5 4

Printed and bound by Von Hoffmann Press

A030514

QS 50

# CONTENTS

► *Atlas of the Human Skeleton* is offered to accompany *Principles of Anatomy and Physiology,* Tenth Edition, by Gerard J. Tortora and Sandra R. Grabowski. The photographs of various components of the skeleton, prepared by Mark Nielsen of the University of Utah, have been carefully selected, oriented, and labeled for maximum effectiveness as both a supplement to the illustrations in the textbook and as a guide for learning the skeletal system in the laboratory.

The order of the photographs in the *Atlas* generally follows that of the illustrations in the textbook. In addition, to assist you further in correlating the *Atlas* photographs with the textbook illustrations, most photos in the *Atlas* are referenced to corresponding textbook figures and a corresponding textbook description of the bones under consideration.

It is hoped that this *Atlas* will provide you with a pedagogically effective presentation of the human skeleton that you can use in the laboratory while working with actual bones or outside of the laboratory in conjunction with your textbook.

Please send me any suggestions for improvement so that I may incorporate them into subsequent editions.

Gerard J. Tortora
Sciences and Technology S229
Bergen Community College
400 Paramus Road
Paramus NJ 07652

## Divisions of the Adult Skeletal System

| Regions of the Skeleton | Number of Bones | Regions of the Skeleton | Number of Bones |
|---|---|---|---|
| **Axial Skeleton** | | **Appendicular Skeleton** | |
| Skull | | Pectoral (shoulder) girdles | |
| *Cranium* | 8 | *Clavicle* | 2 |
| *Face* | 14 | *Scapula* | 2 |
| Hyoid | 1 | Upper limbs (extremities) | |
| Auditory ossicles | 6 | *Humerus* | 2 |
| Vertebral column | 26 | *Ulna* | 2 |
| Thorax | | *Radius* | 2 |
| *Sternum* | 1 | *Carpals* | 16 |
| *Ribs* | 24 | *Metacarpals* | 10 |
| Subtotal = 80 | | *Phalanges* | 28 |
| | | Pelvic (hip) girdle | |
| | | *Hip, pelvic, or coxal bone* | 2 |
| | | Lower limbs (extremities) | |
| | | *Femur* | 2 |
| | | *Fibula* | 2 |
| | | *Tibia* | 2 |
| | | *Patella* | 2 |
| | | *Tarsals* | 14 |
| | | *Metatarsals* | 10 |
| | | *Phalanges* | 28 |
| | | Subtotal = 126 | |
| | | Total = 206 | |

SUPERIOR

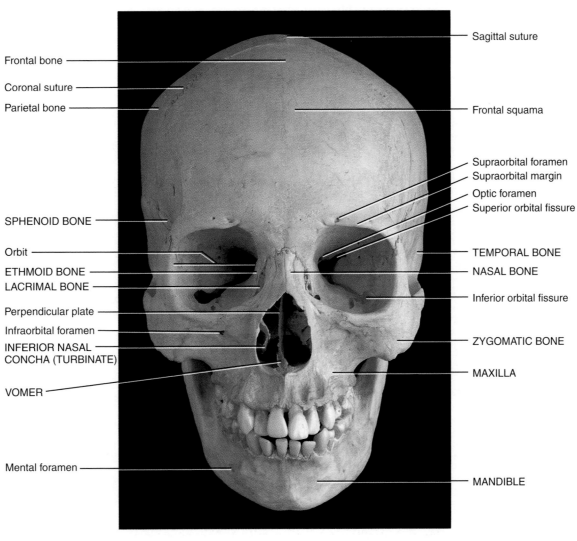

Sagittal suture

Frontal bone

Coronal suture

Parietal bone

Frontal squama

Supraorbital foramen
Supraorbital margin
Optic foramen
Superior orbital fissure

SPHENOID BONE

Orbit

TEMPORAL BONE

ETHMOID BONE

NASAL BONE

LACRIMAL BONE

Inferior orbital fissure

Perpendicular plate

Infraorbital foramen

INFERIOR NASAL
CONCHA (TURBINATE)

ZYGOMATIC BONE

MAXILLA

VOMER

Mental foramen

MANDIBLE

INFERIOR

FIGURE 1   Skull, anterior view

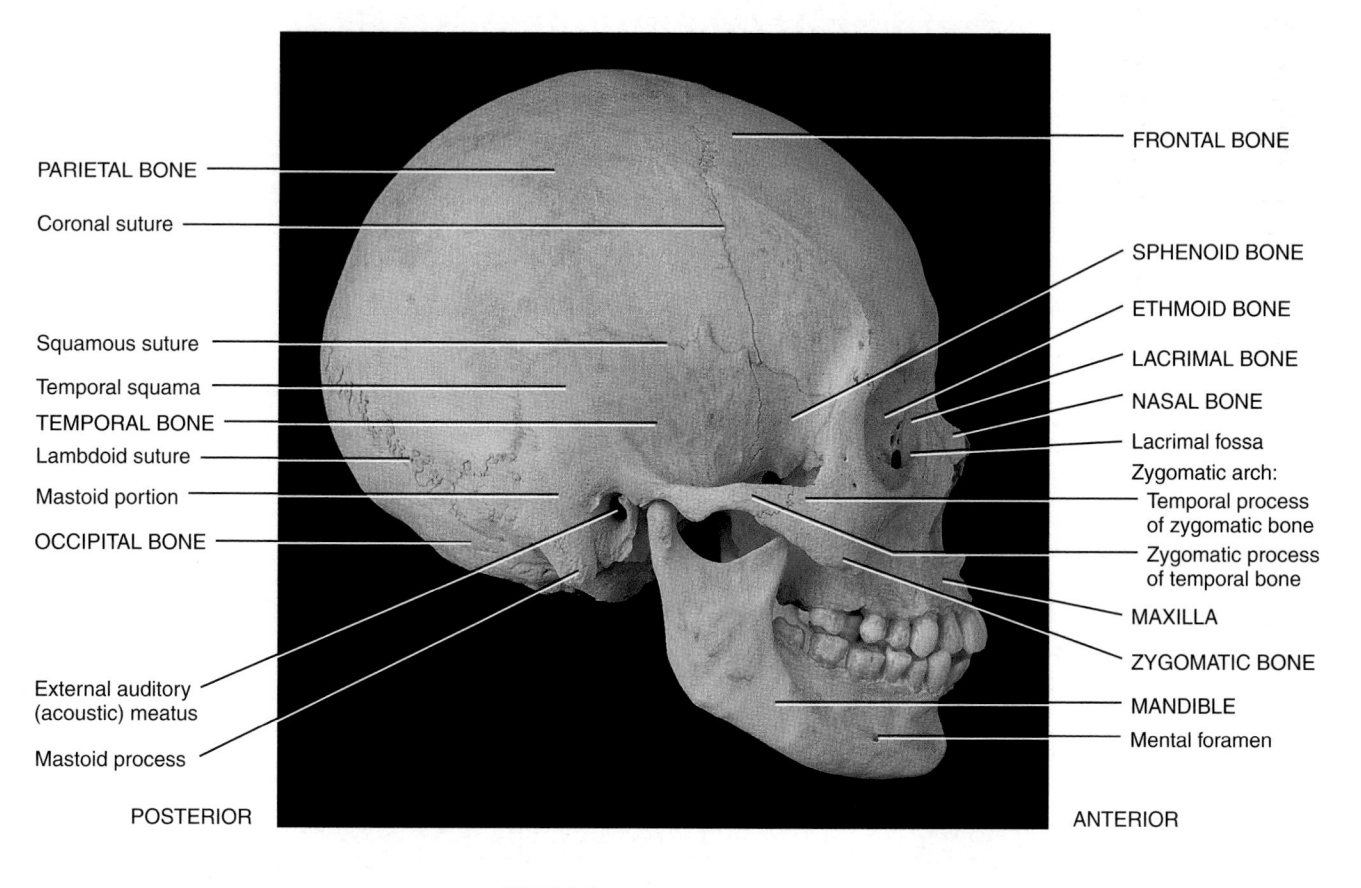

PARIETAL BONE

Coronal suture

Squamous suture

Temporal squama

TEMPORAL BONE

Lambdoid suture

Mastoid portion

OCCIPITAL BONE

External auditory
(acoustic) meatus

Mastoid process

POSTERIOR

FRONTAL BONE

SPHENOID BONE

ETHMOID BONE

LACRIMAL BONE

NASAL BONE

Lacrimal fossa

Zygomatic arch:
  Temporal process
  of zygomatic bone

  Zygomatic process
  of temporal bone

MAXILLA

ZYGOMATIC BONE

MANDIBLE

Mental foramen

ANTERIOR

FIGURE 2   Skull, right lateral view

2

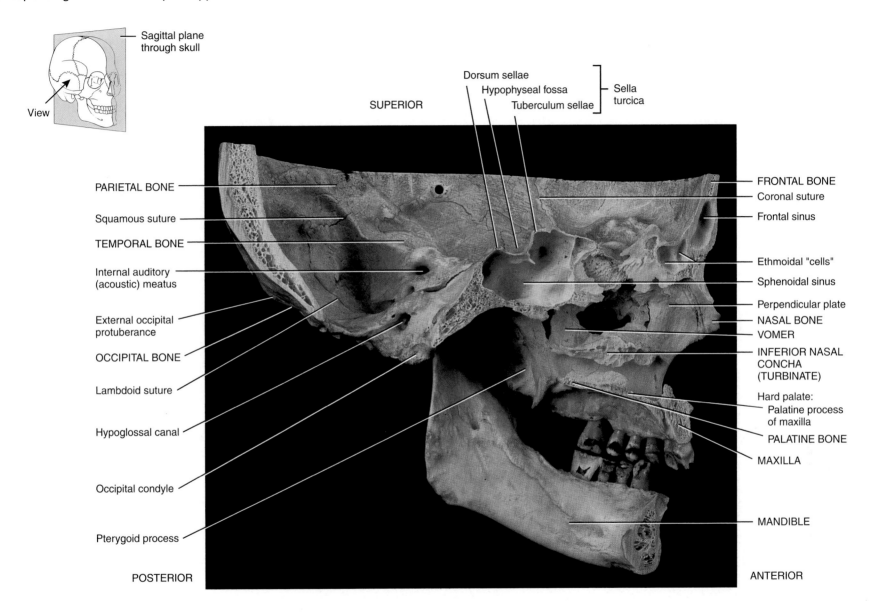

Sagittal plane through skull

View

SUPERIOR

Dorsum sellae
Hypophyseal fossa
Tuberculum sellae
Sella turcica

PARIETAL BONE

Squamous suture

TEMPORAL BONE

Internal auditory (acoustic) meatus

External occipital protuberance

OCCIPITAL BONE

Lambdoid suture

Hypoglossal canal

Occipital condyle

Pterygoid process

FRONTAL BONE
Coronal suture
Frontal sinus

Ethmoidal "cells"

Sphenoidal sinus

Perpendicular plate
NASAL BONE
VOMER

INFERIOR NASAL CONCHA (TURBINATE)

Hard palate:
Palatine process of maxilla
PALATINE BONE

MAXILLA

MANDIBLE

POSTERIOR

ANTERIOR

FIGURE 3   Skull, medial view of sagittal section

3

▶Corresponding textbook figure: 7.6
Corresponding textbook description: pp. 189-202

ANTERIOR

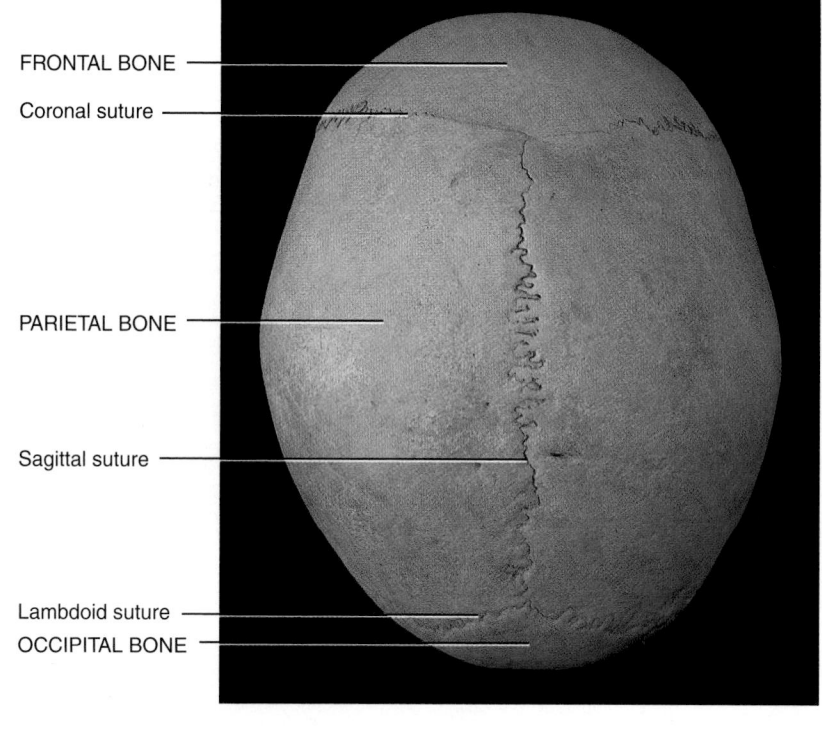

FRONTAL BONE ——————

Coronal suture ——————

PARIETAL BONE ——————

Sagittal suture ——————

Lambdoid suture ——————
OCCIPITAL BONE ——————

POSTERIOR

FIGURE 5   Skull, superior view

SUPERIOR

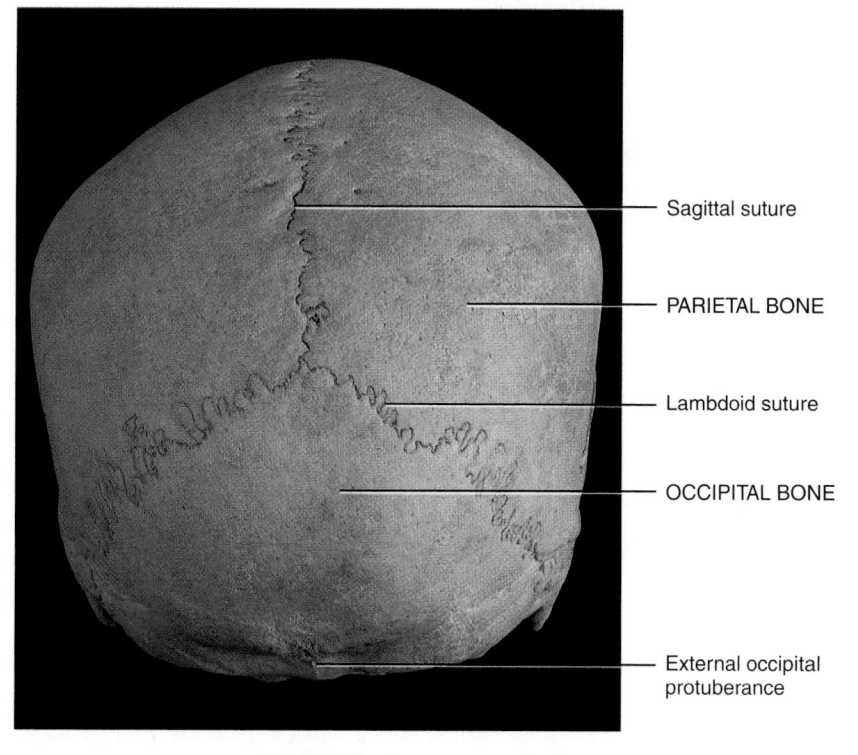

—— Sagittal suture

—— PARIETAL BONE

—— Lambdoid suture

—— OCCIPITAL BONE

—— External occipital
protuberance

INFERIOR

FIGURE 4   Skull, posterior view

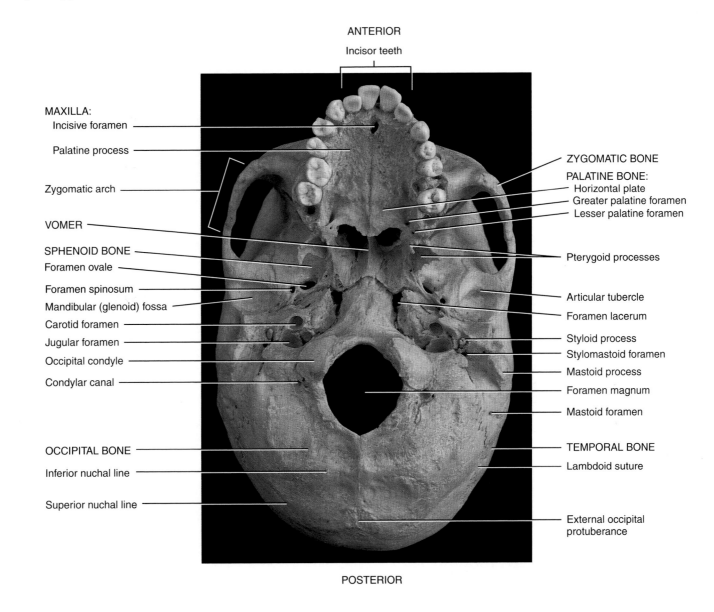

ANTERIOR

Incisor teeth

View

MAXILLA:
  Incisive foramen

  Palatine process

  Zygomatic arch

VOMER

SPHENOID BONE
  Foramen ovale

  Foramen spinosum

  Mandibular (glenoid) fossa

  Carotid foramen

  Jugular foramen

  Occipital condyle

  Condylar canal

OCCIPITAL BONE

  Inferior nuchal line

  Superior nuchal line

ZYGOMATIC BONE

PALATINE BONE:
  Horizontal plate
  Greater palatine foramen
  Lesser palatine foramen

Pterygoid processes

Articular tubercle

Foramen lacerum

Styloid process

Stylomastoid foramen

Mastoid process

Foramen magnum

Mastoid foramen

TEMPORAL BONE

Lambdoid suture

External occipital
protuberance

POSTERIOR

FIGURE 6   Skull, inferior view

5

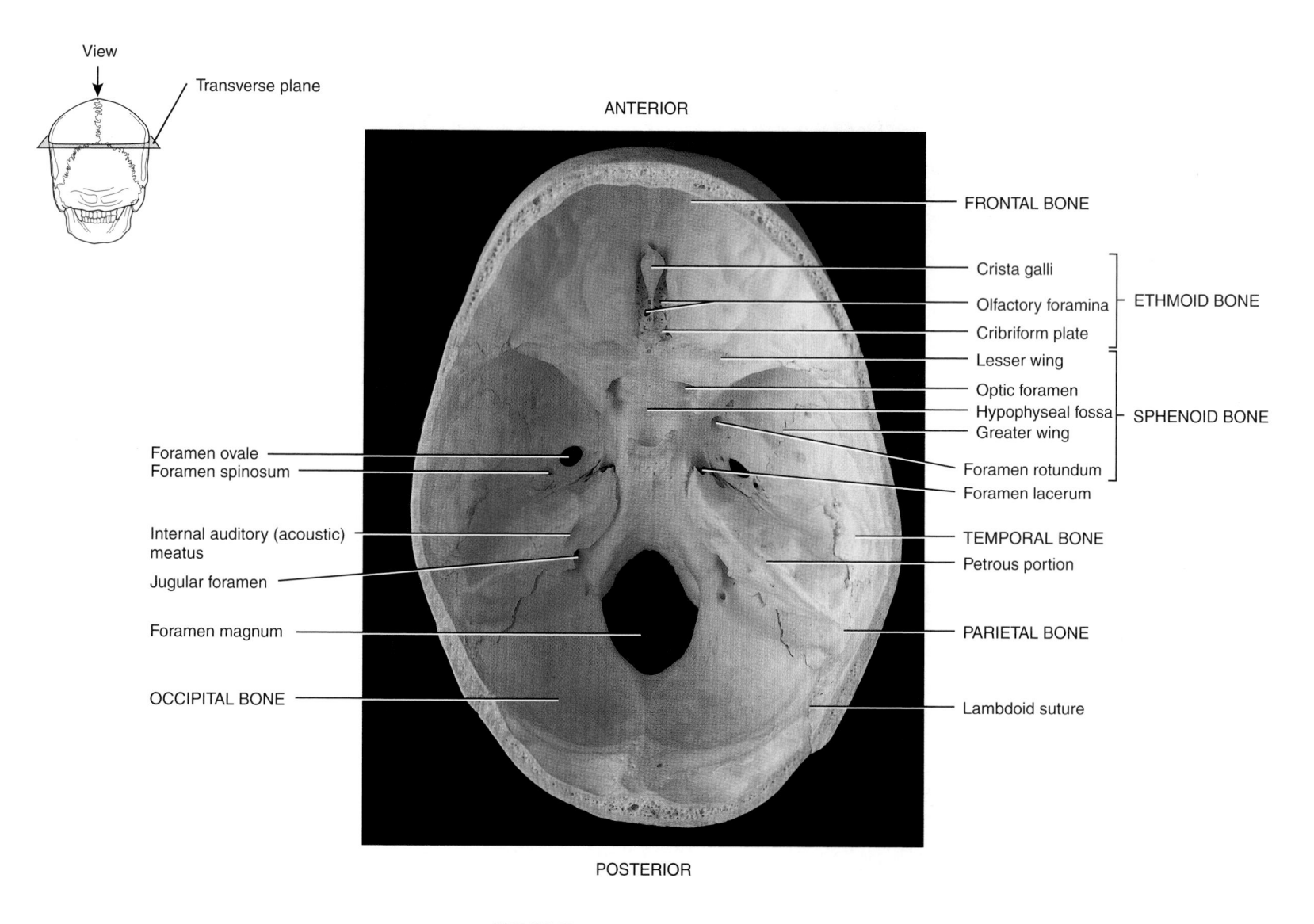

View

Transverse plane

ANTERIOR

FRONTAL BONE

Crista galli

Olfactory foramina — ETHMOID BONE

Cribriform plate

Lesser wing

Optic foramen

Hypophyseal fossa — SPHENOID BONE

Greater wing

Foramen ovale

Foramen spinosum

Foramen rotundum

Foramen lacerum

Internal auditory (acoustic) meatus

TEMPORAL BONE

Petrous portion

Jugular foramen

Foramen magnum

PARIETAL BONE

OCCIPITAL BONE

Lambdoid suture

POSTERIOR

FIGURE 7  Floor of cranium, superior view

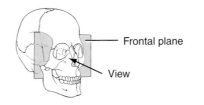

Frontal plane

View

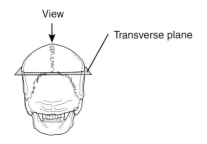

View

Transverse plane

ANTERIOR

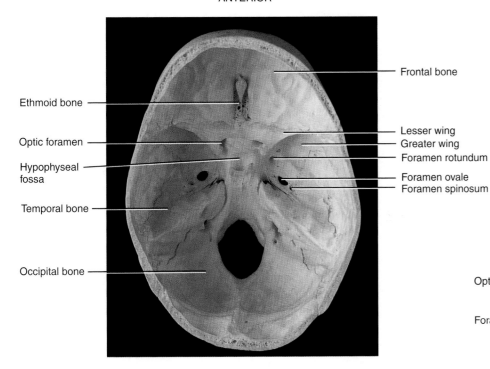

Ethmoid bone

Optic foramen

Hypophyseal
fossa

Temporal bone

Occipital bone

Frontal bone

Lesser wing
Greater wing
Foramen rotundum

Foramen ovale
Foramen spinosum

FIGURE 8   Sphenoid bone, superior view

SUPERIOR

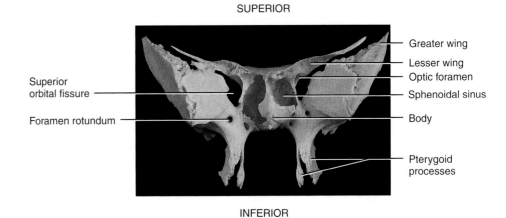

Superior
orbital fissure

Foramen rotundum

Greater wing
Lesser wing
Optic foramen
Sphenoidal sinus

Body

Pterygoid
processes

INFERIOR

FIGURE 9   Sphenoid bone, anterior view

SUPERIOR

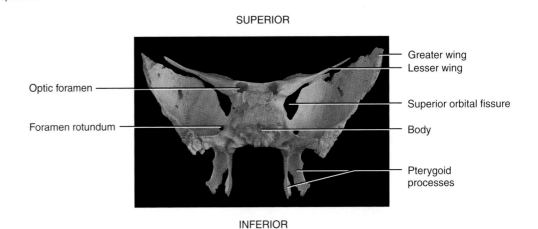

Optic foramen

Foramen rotundum

Greater wing
Lesser wing

Superior orbital fissure

Body

Pterygoid
processes

INFERIOR

FIGURE 10   Sphenoid bone, posterior view

7

Sagittal plane

View

SUPERIOR

Hypophyseal fossa

Sphenoidal sinus

Inferior nasal concha (turbinate)

Mandible

Frontal sinus

Ethmoidal "cells"

Perpendicular plate

Nasal bone

Maxilla

POSTERIOR

ANTERIOR

FIGURE 11    Ethmoid bone, medial view of sagittal section

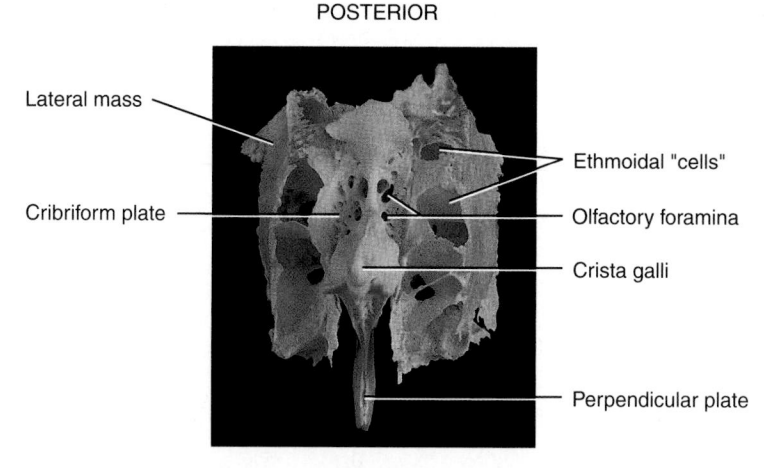

POSTERIOR

Lateral mass

Cribriform plate

Ethmoidal "cells"

Olfactory foramina

Crista galli

Perpendicular plate

ANTERIOR

FIGURE 12    Ethmoid bone, superior view

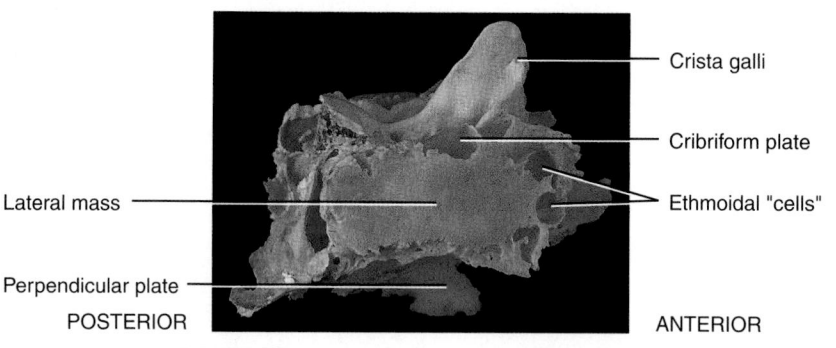

Crista galli

Cribriform plate

Ethmoidal "cells"

Lateral mass

Perpendicular plate

POSTERIOR

ANTERIOR

FIGURE 13    Ethmoid bone, right lateral view

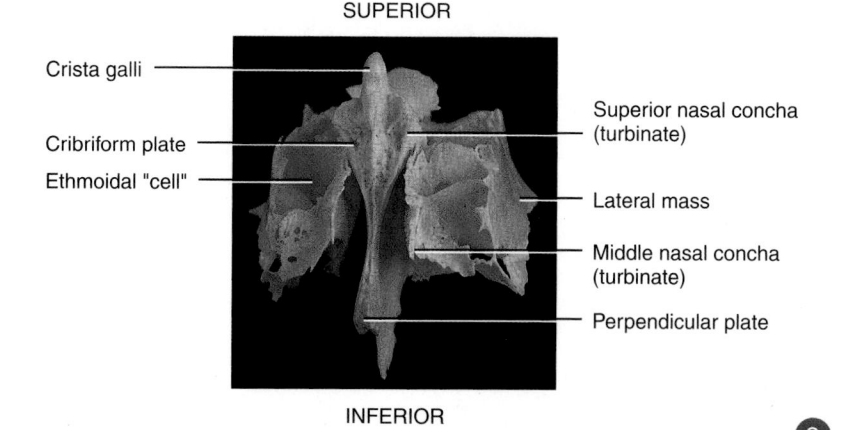

SUPERIOR

Crista galli

Cribriform plate

Ethmoidal "cell"

Superior nasal concha (turbinate)

Lateral mass

Middle nasal concha (turbinate)

Perpendicular plate

INFERIOR

FIGURE 14    Ethmoid bone, anterior view

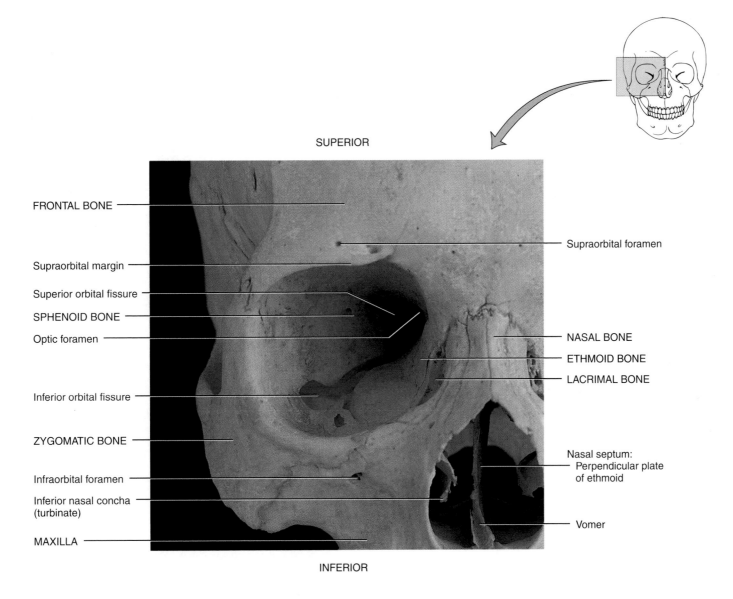

SUPERIOR

FRONTAL BONE

Supraorbital foramen

Supraorbital margin

Superior orbital fissure

SPHENOID BONE

Optic foramen

NASAL BONE

ETHMOID BONE

LACRIMAL BONE

Inferior orbital fissure

ZYGOMATIC BONE

Nasal septum:
  Perpendicular plate
  of ethmoid

Infraorbital foramen

Inferior nasal concha
(turbinate)

Vomer

MAXILLA

INFERIOR

FIGURE 15   Right orbit, anterior view

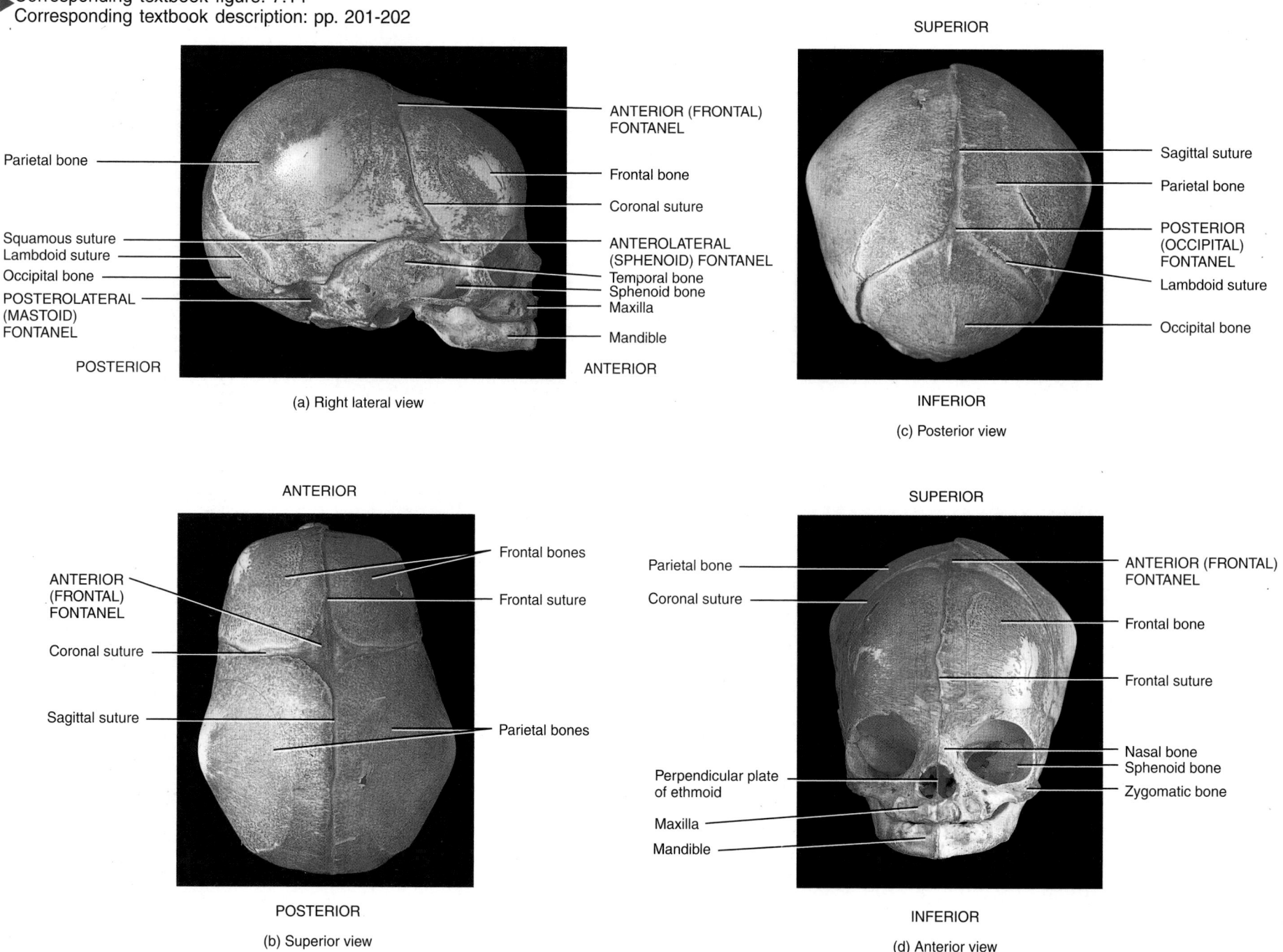

(a) Right lateral view

(b) Superior view

(c) Posterior view

(d) Anterior view

FIGURE 16  Fetal skull, fontanels

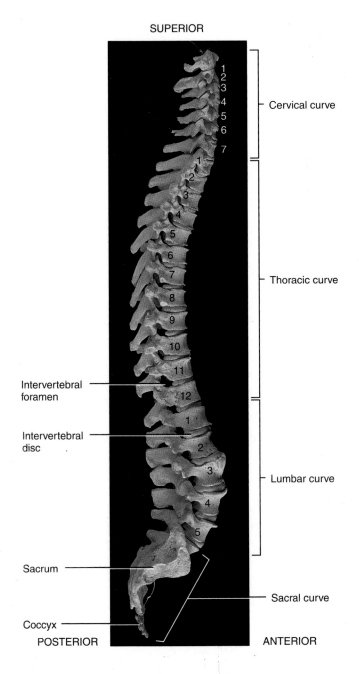

SUPERIOR

- Cervical curve
  - 1
  - 2
  - 3
  - 4
  - 5
  - 6
  - 7
- Thoracic curve
  - 1
  - 2
  - 3
  - 4
  - 5
  - 6
  - 7
  - 8
  - 9
  - 10
  - 11
  - 12
- Lumbar curve
  - 1
  - 2
  - 3
  - 4
  - 5
- Sacral curve

Intervertebral foramen

Intervertebral disc

Sacrum

Coccyx

POSTERIOR

ANTERIOR

FIGURE 17　Vertebral column, right lateral view

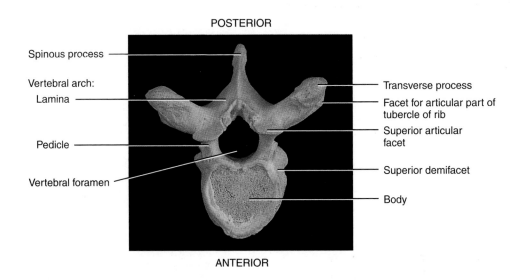

POSTERIOR

Spinous process

Vertebral arch:
　Lamina

Pedicle

Vertebral foramen

Transverse process

Facet for articular part of tubercle of rib

Superior articular facet

Superior demifacet

Body

ANTERIOR

FIGURE 18　A typical (thoracic) vertebra, superior view

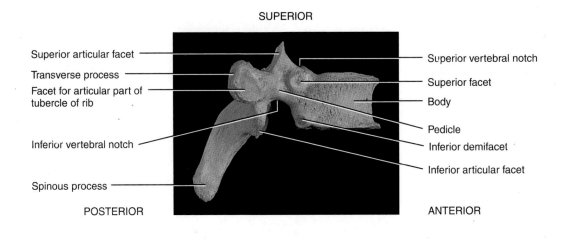

SUPERIOR

Superior articular facet

Transverse process

Facet for articular part of tubercle of rib

Inferior vertebral notch

Spinous process

Superior vertebral notch

Superior facet

Body

Pedicle

Inferior demifacet

Inferior articular facet

POSTERIOR

ANTERIOR

FIGURE 19　A typical (thoracic) vertebra, right lateral view

Corresponding textbook figure: 7.18
Corresponding textbook description: pp. 205-206

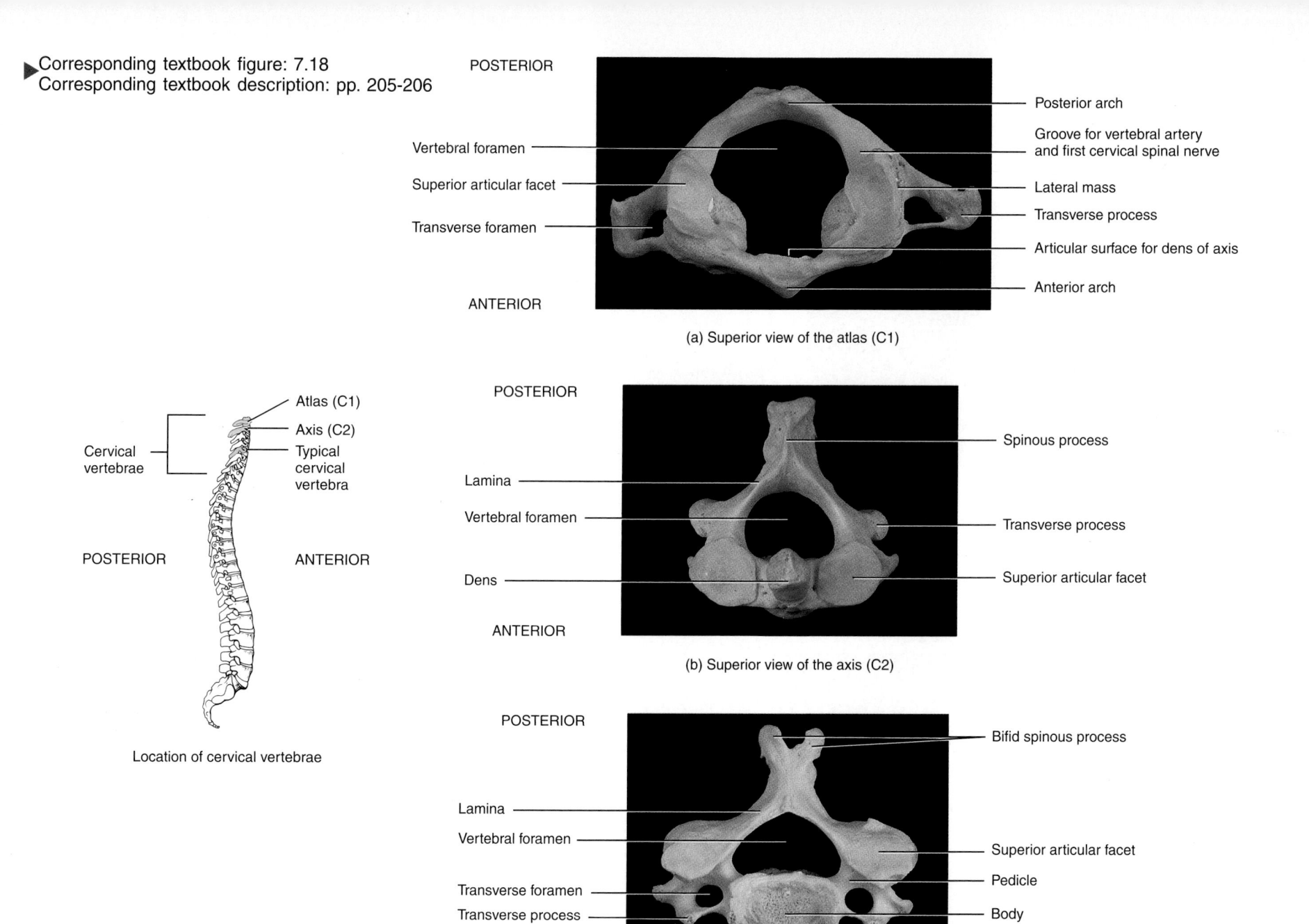

POSTERIOR

Vertebral foramen

Superior articular facet

Transverse foramen

ANTERIOR

Posterior arch

Groove for vertebral artery and first cervical spinal nerve

Lateral mass

Transverse process

Articular surface for dens of axis

Anterior arch

(a) Superior view of the atlas (C1)

Atlas (C1)

Axis (C2)

Typical cervical vertebra

Cervical vertebrae

POSTERIOR          ANTERIOR

Location of cervical vertebrae

POSTERIOR

Lamina

Vertebral foramen

Dens

ANTERIOR

Spinous process

Transverse process

Superior articular facet

(b) Superior view of the axis (C2)

POSTERIOR

Lamina

Vertebral foramen

Transverse foramen

Transverse process

ANTERIOR

Bifid spinous process

Superior articular facet

Pedicle

Body

(c) Superior view of a typical cervical vertebra

FIGURE 20   Cervical vertebrae

12

Corresponding textbook figure: 7.20
Corresponding textbook description: p. 207

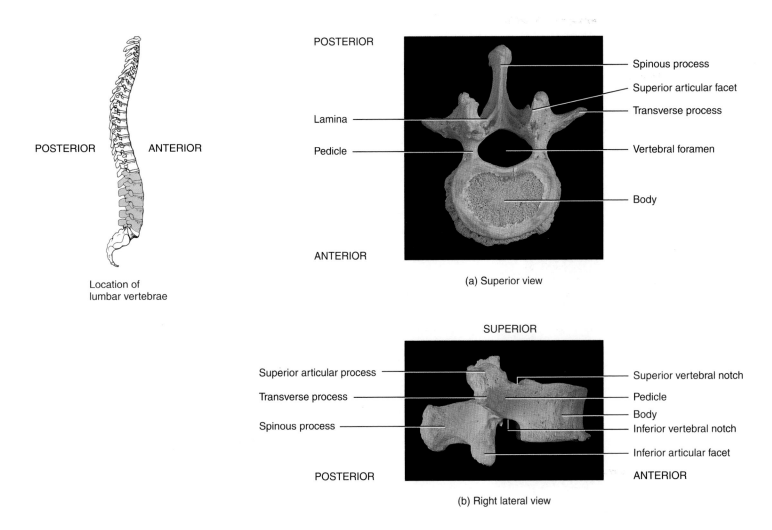

POSTERIOR

POSTERIOR          ANTERIOR

Location of
lumbar vertebrae

POSTERIOR

Lamina

Pedicle

ANTERIOR

Spinous process

Superior articular facet

Transverse process

Vertebral foramen

Body

(a) Superior view

SUPERIOR

Superior articular process

Transverse process

Spinous process

Superior vertebral notch

Pedicle

Body

Inferior vertebral notch

Inferior articular facet

POSTERIOR          ANTERIOR

(b) Right lateral view

FIGURE 21   Lumbar vertebrae

13

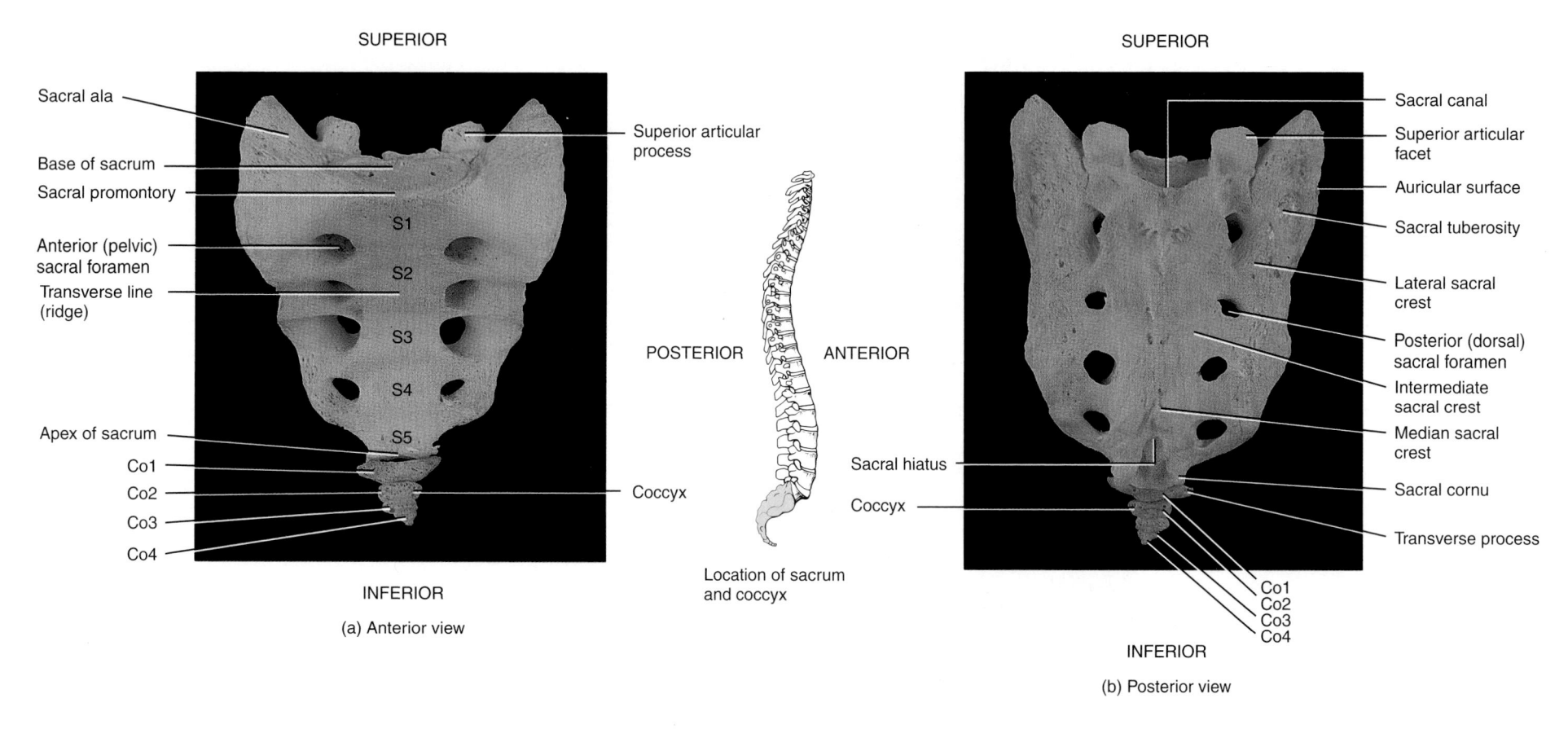

SUPERIOR

Sacral ala

Base of sacrum

Sacral promontory

Superior articular
process

S1

Anterior (pelvic)
sacral foramen

S2

Transverse line
(ridge)

S3

S4

Apex of sacrum

S5

Co1

Co2

Coccyx

Co3

Co4

INFERIOR

(a) Anterior view

POSTERIOR          ANTERIOR

Location of sacrum
and coccyx

SUPERIOR

Sacral canal

Superior articular
facet

Auricular surface

Sacral tuberosity

Lateral sacral
crest

Posterior (dorsal)
sacral foramen

Intermediate
sacral crest

Median sacral
crest

Sacral hiatus

Sacral cornu

Coccyx

Transverse process

Co1
Co2
Co3
Co4

INFERIOR

(b) Posterior view

FIGURE 22   Sacrum and coccyx

▶Corresponding textbook figure: 7.22
Corresponding textbook description: pp. 211-212

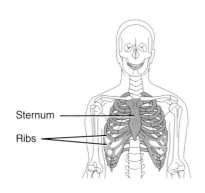

Sternum

Ribs

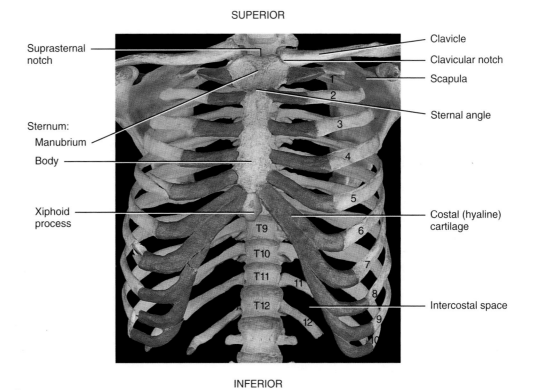

SUPERIOR

Suprasternal notch

Clavicle

Clavicular notch

Scapula

1

2

Sternal angle

3

Sternum:
Manubrium

4

Body

5

Xiphoid process

Costal (hyaline) cartilage

T9

6

T10

7

T11

11

8

T12

Intercostal space

12

9

10

INFERIOR

FIGURE 23   Skeleton of thorax, anterior view

SUPERIOR

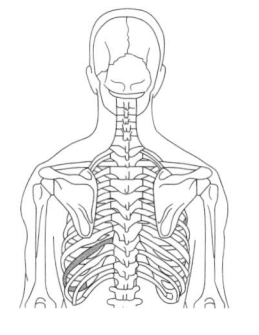

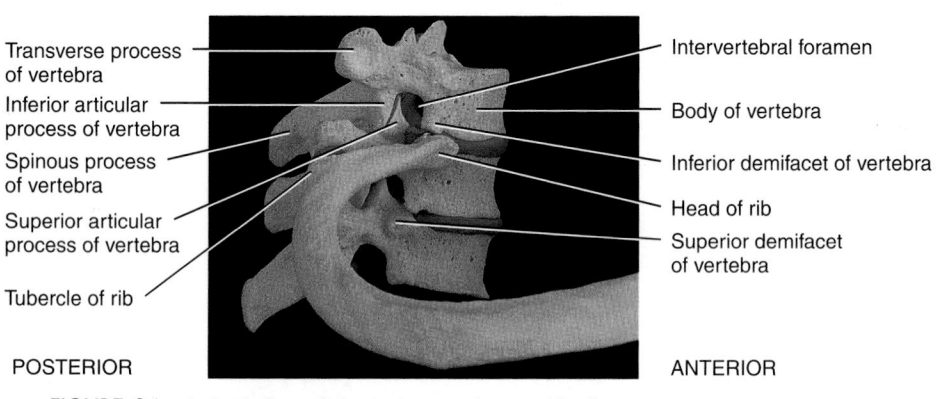

Transverse process of vertebra

Inferior articular process of vertebra

Spinous process of vertebra

Superior articular process of vertebra

Tubercle of rib

Intervertebral foramen

Body of vertebra

Inferior demifacet of vertebra

Head of rib

Superior demifacet of vertebra

POSTERIOR

ANTERIOR

FIGURE 24   Articulation of thoracic vertebrae with rib, right lateral view

POSTERIOR

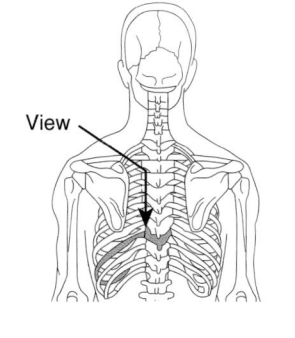

View

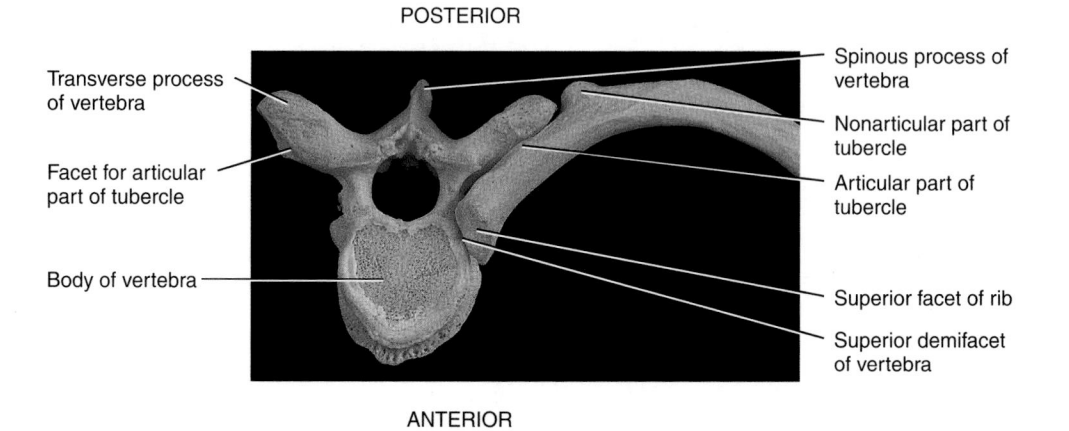

Transverse process of vertebra

Facet for articular part of tubercle

Body of vertebra

Spinous process of vertebra

Nonarticular part of tubercle

Articular part of tubercle

Superior facet of rib

Superior demifacet of vertebra

ANTERIOR

FIGURE 25   Articulation of thoracic vertebra with rib, superior view

Corresponding textbook figure: 8.3
Corresponding textbook description: p. 220

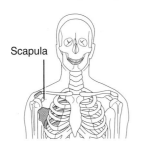

Scapula

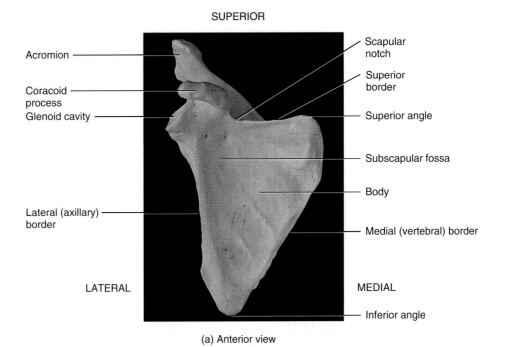

SUPERIOR

Acromion

Coracoid process

Glenoid cavity

Scapular notch

Superior border

Superior angle

Subscapular fossa

Body

Lateral (axillary) border

Medial (vertebral) border

LATERAL

MEDIAL

Inferior angle

(a) Anterior view

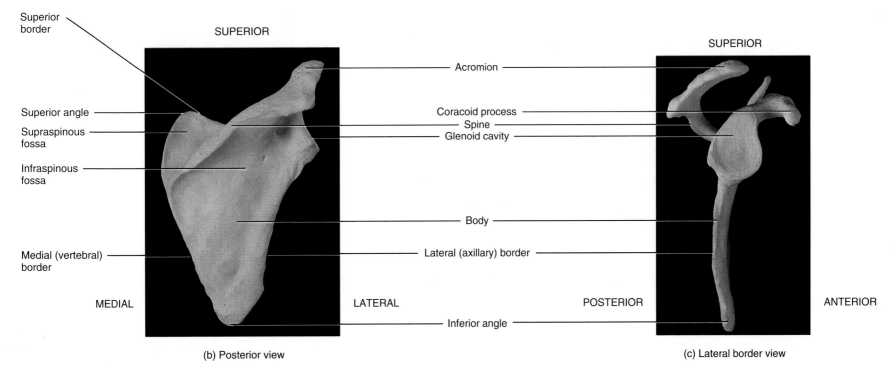

Superior border

SUPERIOR

Acromion

Superior angle

Coracoid process

Supraspinous fossa

Spine

Glenoid cavity

Infraspinous fossa

Medial (vertebral) border

Body

Lateral (axillary) border

MEDIAL

LATERAL

POSTERIOR

ANTERIOR

Inferior angle

(b) Posterior view

SUPERIOR

(c) Lateral border view

FIGURE 26   Right scapula

Corresponding textbook figure: 8.5
Corresponding textbook description: p. 222

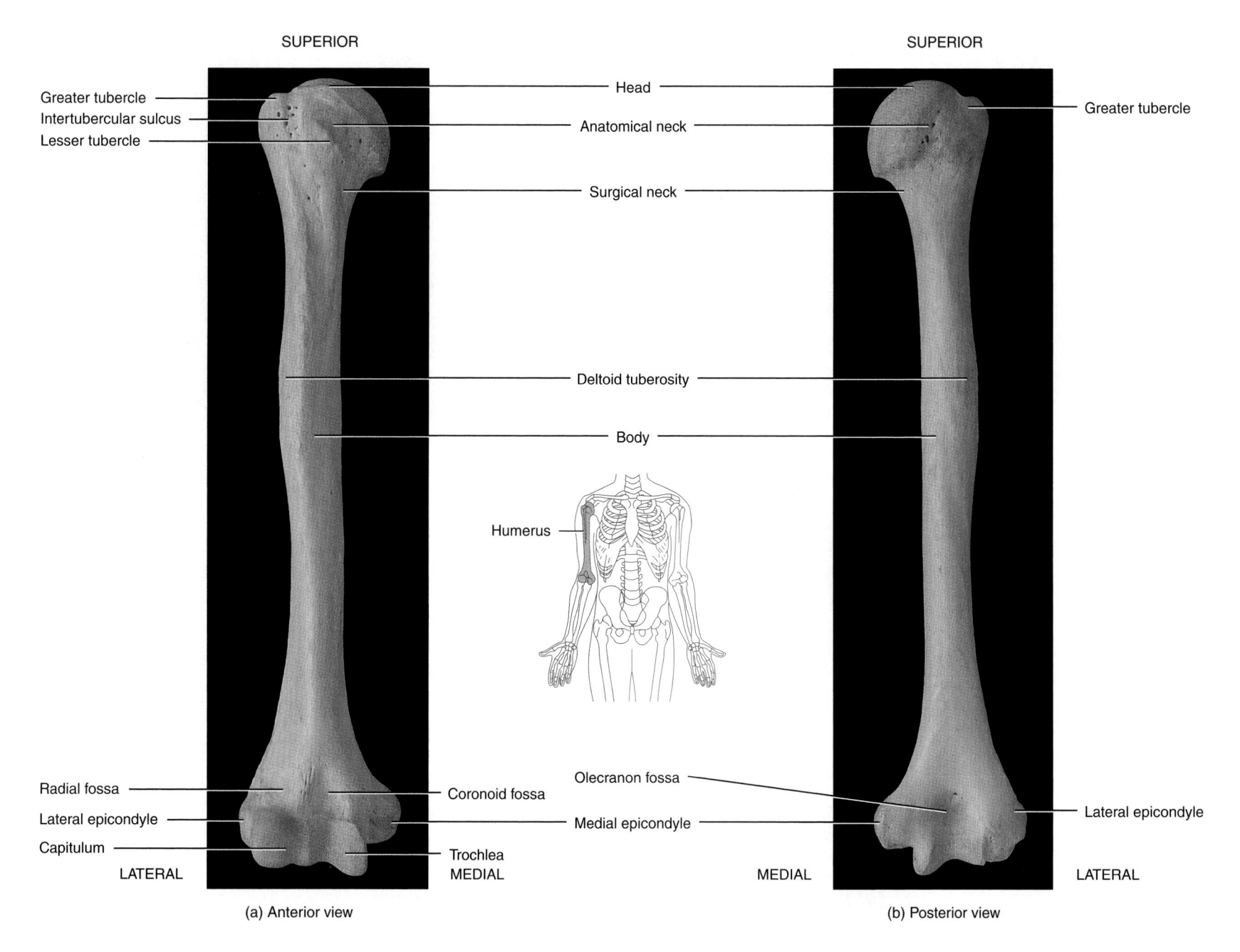

(a) Anterior view

(b) Posterior view

FIGURE 27  Right humerus

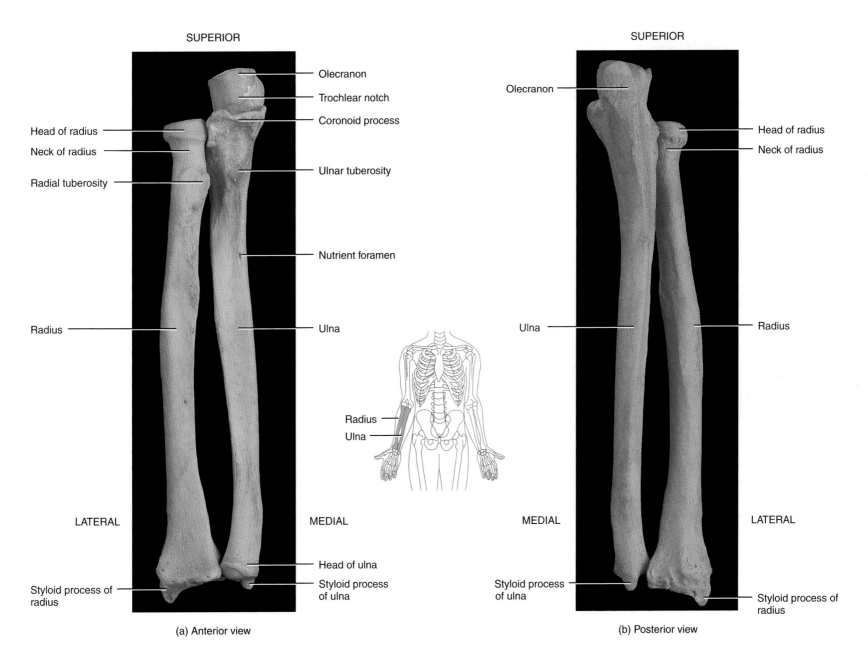

SUPERIOR

Olecranon

Trochlear notch

Coronoid process

Head of radius

Neck of radius

Radial tuberosity

Ulnar tuberosity

Nutrient foramen

Radius

Ulna

Radius

Ulna

LATERAL

MEDIAL

Head of ulna

Styloid process of
radius

Styloid process
of ulna

(a) Anterior view

SUPERIOR

Olecranon

Head of radius

Neck of radius

Ulna

Radius

MEDIAL

LATERAL

Styloid process
of ulna

Styloid process of
radius

(b) Posterior view

FIGURE 28   Right ulna and radius

19

Corresponding textbook figure: 8.8
Corresponding textbook description: pp. 226-227

Carpals
Metacarpals
Phalanges


SUPERIOR
CARPALS:
SCAPHOID
TRAPEZIUM
TRAPEZOID
CARPALS:
LUNATE
TRIQUETRUM
PISIFORM
CAPITATE
HAMATE
I
II
III
IV
V
METACARPALS
Base
Shaft
Head
PHALANGES:
Proximal
Middle
Distal
Thumb
Index finger
Middle finger
Ring finger
Little finger
LATERAL
MEDIAL


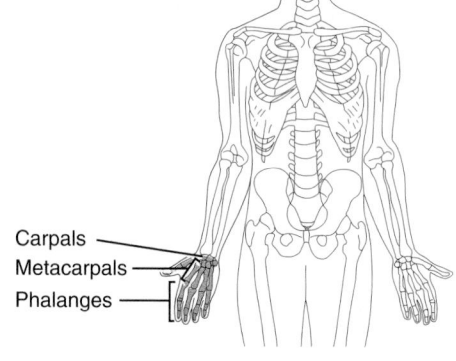

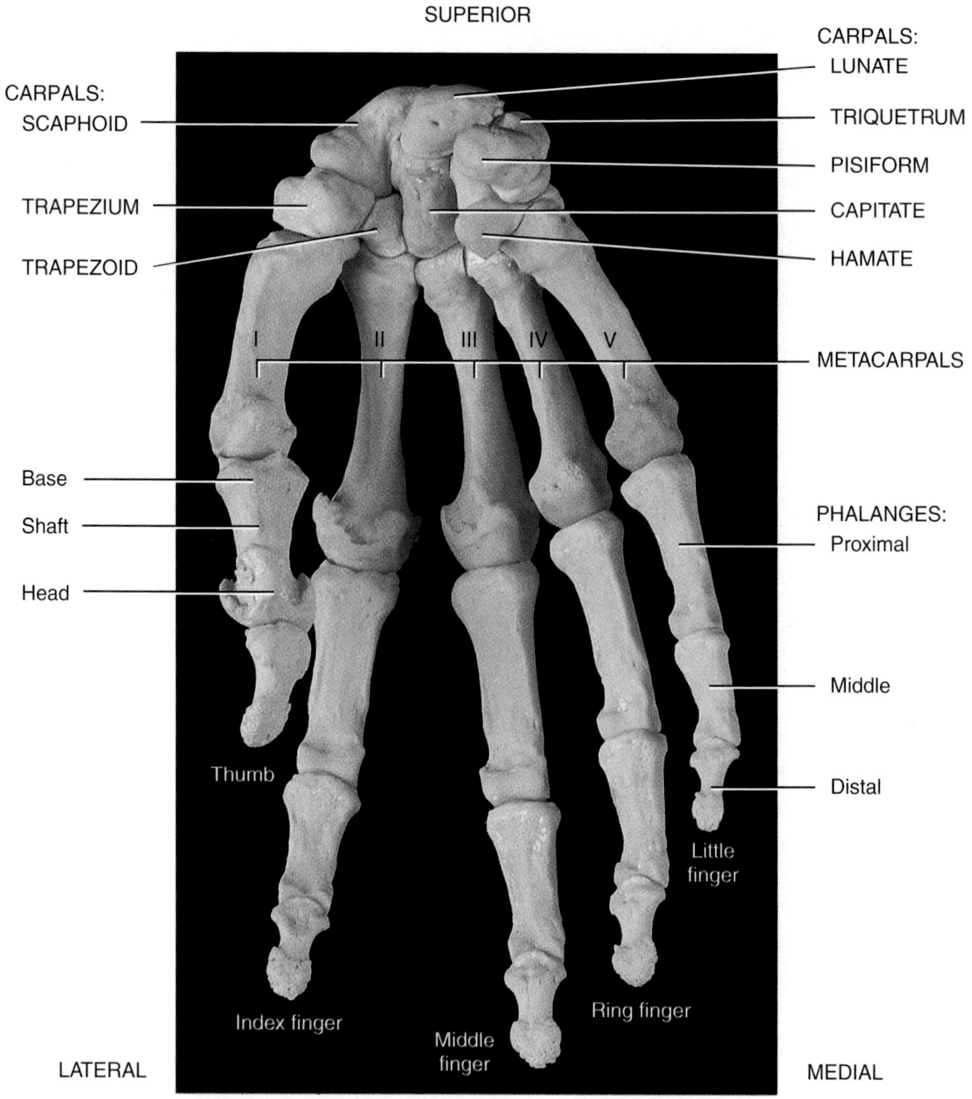

FIGURE 29 Right hand and wrist, anterior view

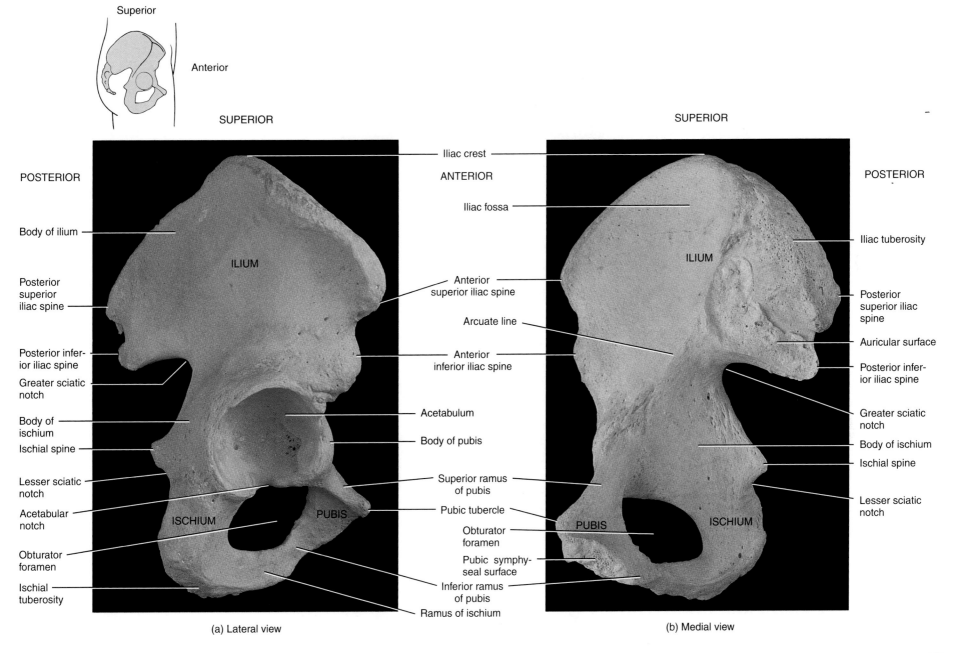

Superior

Anterior

**SUPERIOR**

**SUPERIOR**

Iliac crest

**POSTERIOR**

**ANTERIOR**

**POSTERIOR**

Iliac fossa

Body of ilium

Iliac tuberosity

ILIUM

ILIUM

Posterior
superior
iliac spine

Anterior
superior iliac spine

Posterior
superior iliac
spine

Arcuate line

Posterior infer-
ior iliac spine

Anterior
inferior iliac spine

Auricular surface

Greater sciatic
notch

Posterior infer-
ior iliac spine

Body of
ischium

Acetabulum

Greater sciatic
notch

Ischial spine

Body of pubis

Body of ischium

Ischial spine

Lesser sciatic
notch

Superior ramus
of pubis

Acetabular
notch

Pubic tubercle

Lesser sciatic
notch

ISCHIUM

PUBIS

Obturator
foramen

PUBIS

ISCHIUM

Obturator
foramen

Pubic symphy-
seal surface

Ischial
tuberosity

Inferior ramus
of pubis

Ramus of ischium

(a) Lateral view

(b) Medial view

FIGURE 30   Right hip bone

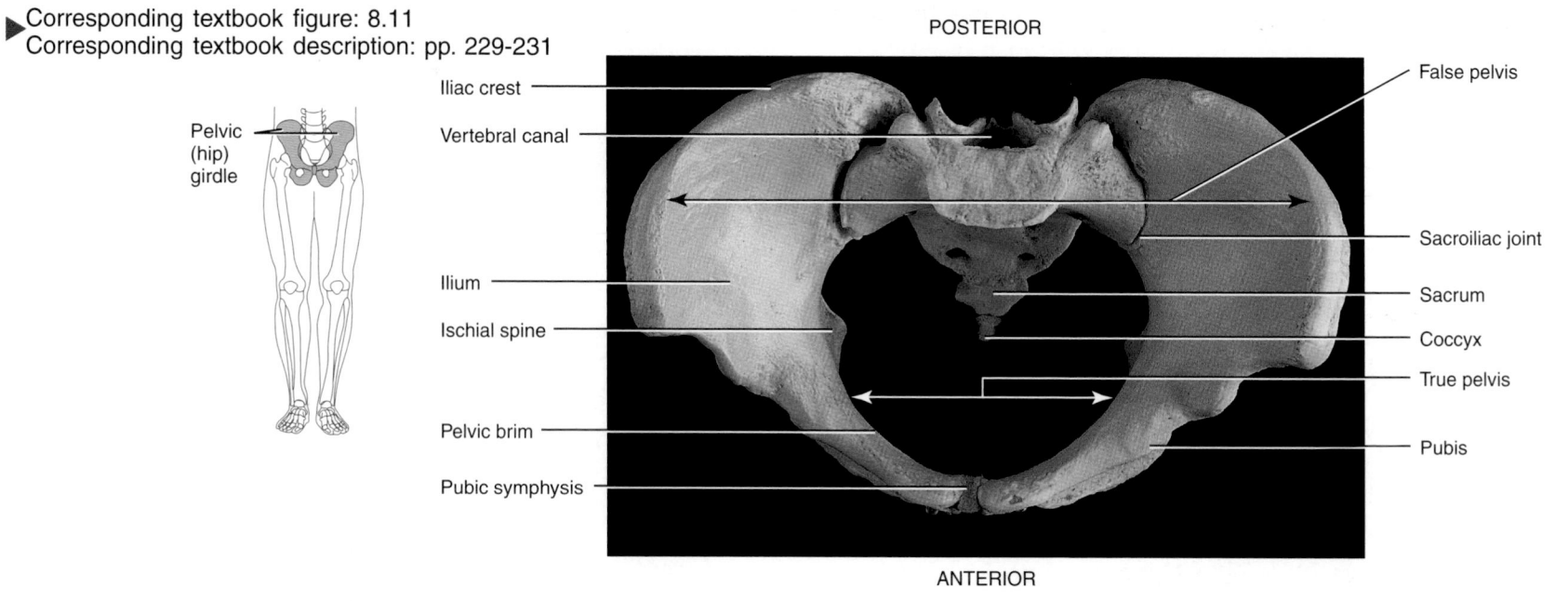

Pelvic (hip) girdle

POSTERIOR

Iliac crest

Vertebral canal

Ilium

Ischial spine

Pelvic brim

Pubic symphysis

False pelvis

Sacroiliac joint

Sacrum

Coccyx

True pelvis

Pubis

ANTERIOR

(a) Female pelvis, superior view

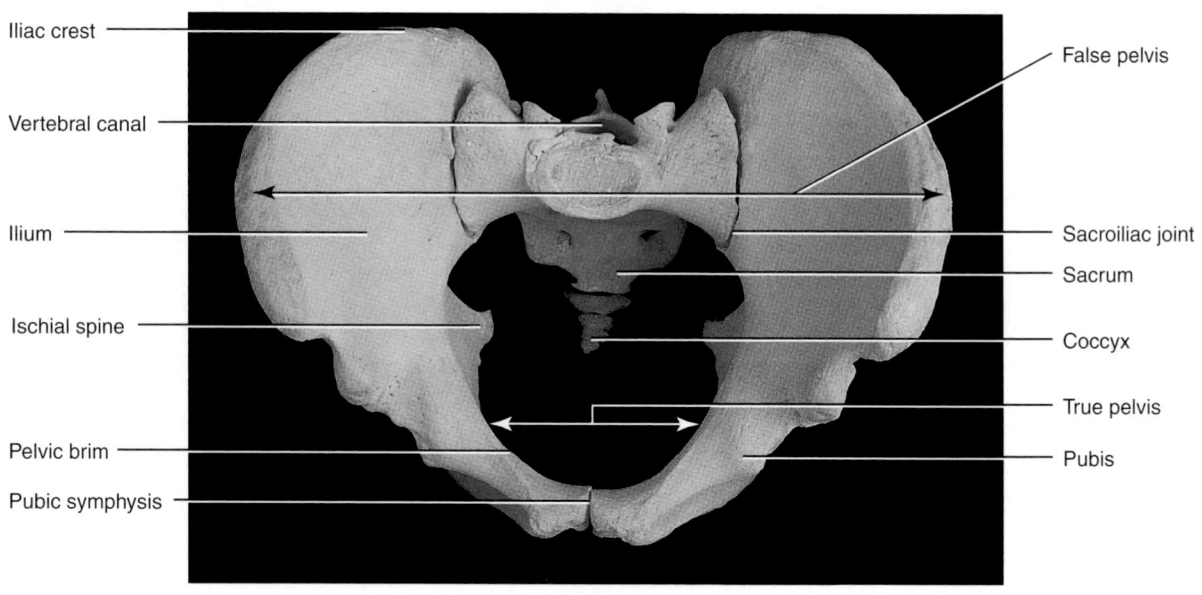

POSTERIOR

Iliac crest

Vertebral canal

Ilium

Ischial spine

Pelvic brim

Pubic symphysis

False pelvis

Sacroiliac joint

Sacrum

Coccyx

True pelvis

Pubis

ANTERIOR

(b) Male pelvis, superior view

FIGURE 31   Pelvis

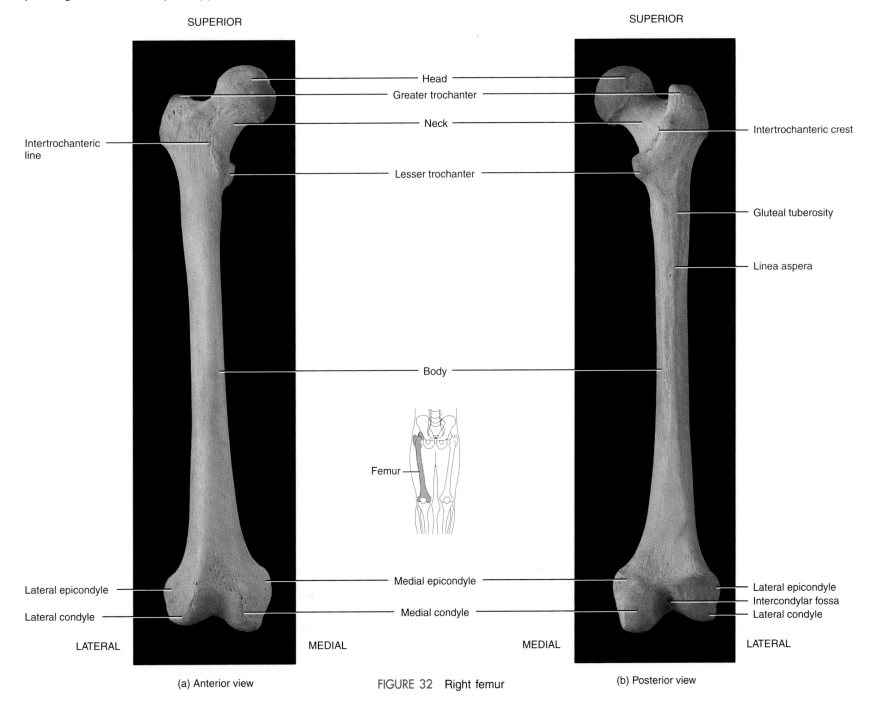

SUPERIOR

SUPERIOR

Head

Greater trochanter

Neck

Intertrochanteric crest

Intertrochanteric
line

Lesser trochanter

Gluteal tuberosity

Linea aspera

Body

Femur

Lateral epicondyle

Medial epicondyle

Lateral epicondyle
Intercondylar fossa

Lateral condyle

Medial condyle

Lateral condyle

LATERAL

MEDIAL

MEDIAL

LATERAL

(a) Anterior view

FIGURE 32   Right femur

(b) Posterior view

▶Corresponding textbook figure: 8.15
Corresponding textbook description: pp. 234-235

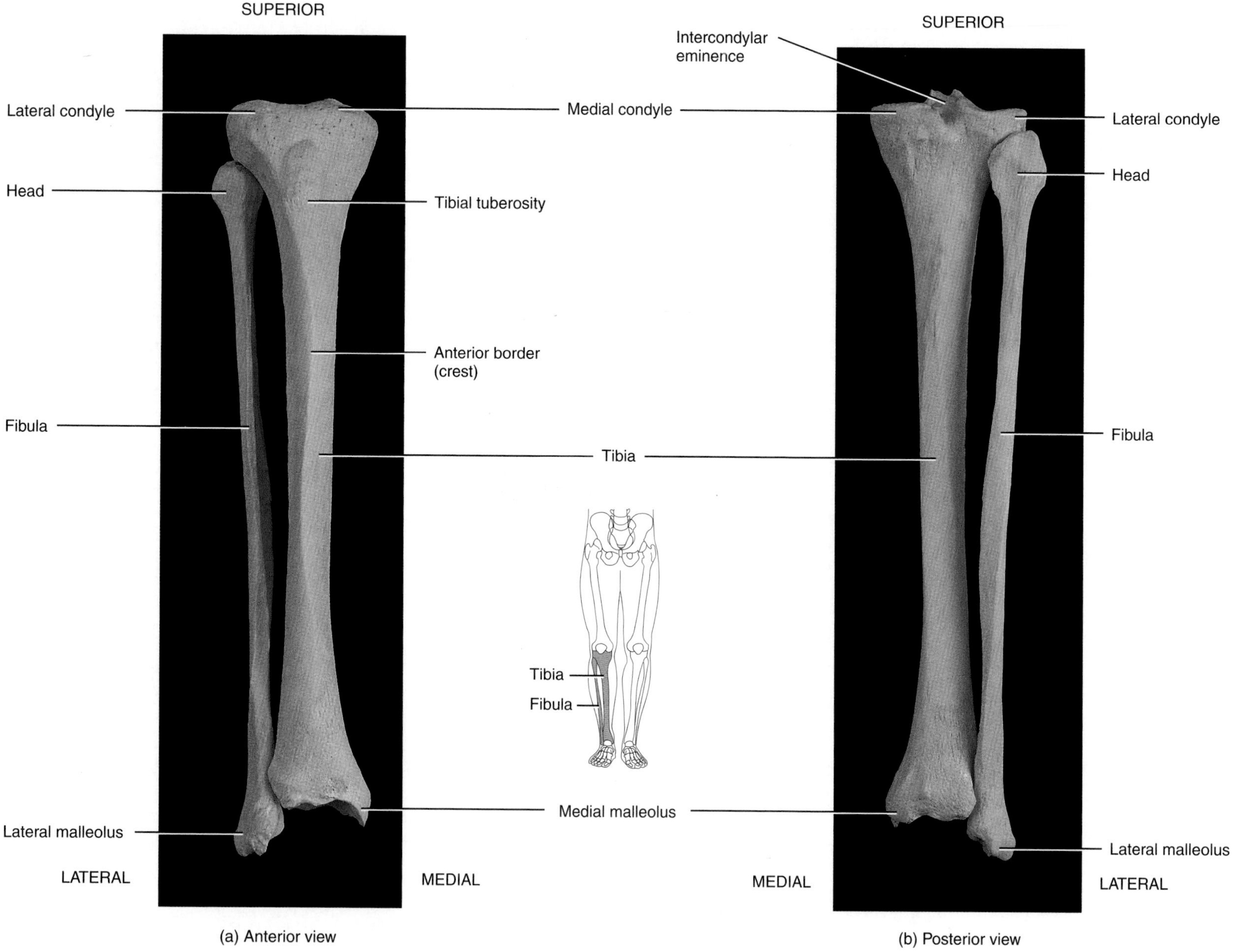
SUPERIOR
Lateral condyle
Head
Fibula
Lateral malleolus
LATERAL
MEDIAL
Medial condyle
Tibial tuberosity
Anterior border
(crest)
Tibia
Medial malleolus
Tibia
Fibula
(a) Anterior view
SUPERIOR
Intercondylar
eminence
Lateral condyle
Head
Fibula
Lateral malleolus
MEDIAL
LATERAL
(b) Posterior view

FIGURE 33 Right tibia and fibula

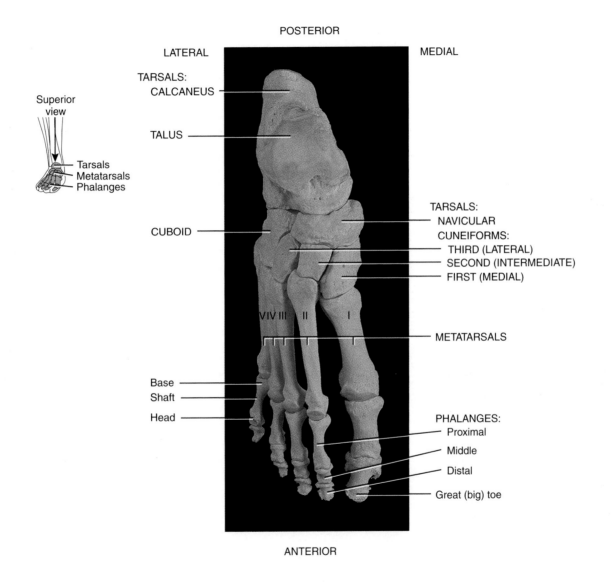

POSTERIOR

LATERAL                                    MEDIAL

TARSALS:
  CALCANEUS

TALUS

Superior
view

  Tarsals
  Metatarsals
  Phalanges

TARSALS:
  NAVICULAR
  CUNEIFORMS:
    THIRD (LATERAL)
    SECOND (INTERMEDIATE)
    FIRST (MEDIAL)

CUBOID

V IV III  II    I

METATARSALS

Base
Shaft
Head

PHALANGES:
  Proximal

  Middle

  Distal

  Great (big) toe

ANTERIOR

FIGURE 34   Right ankle and foot, superior view

Corresponding textbook figure: 7.1
Corresponding textbook description: p. 186

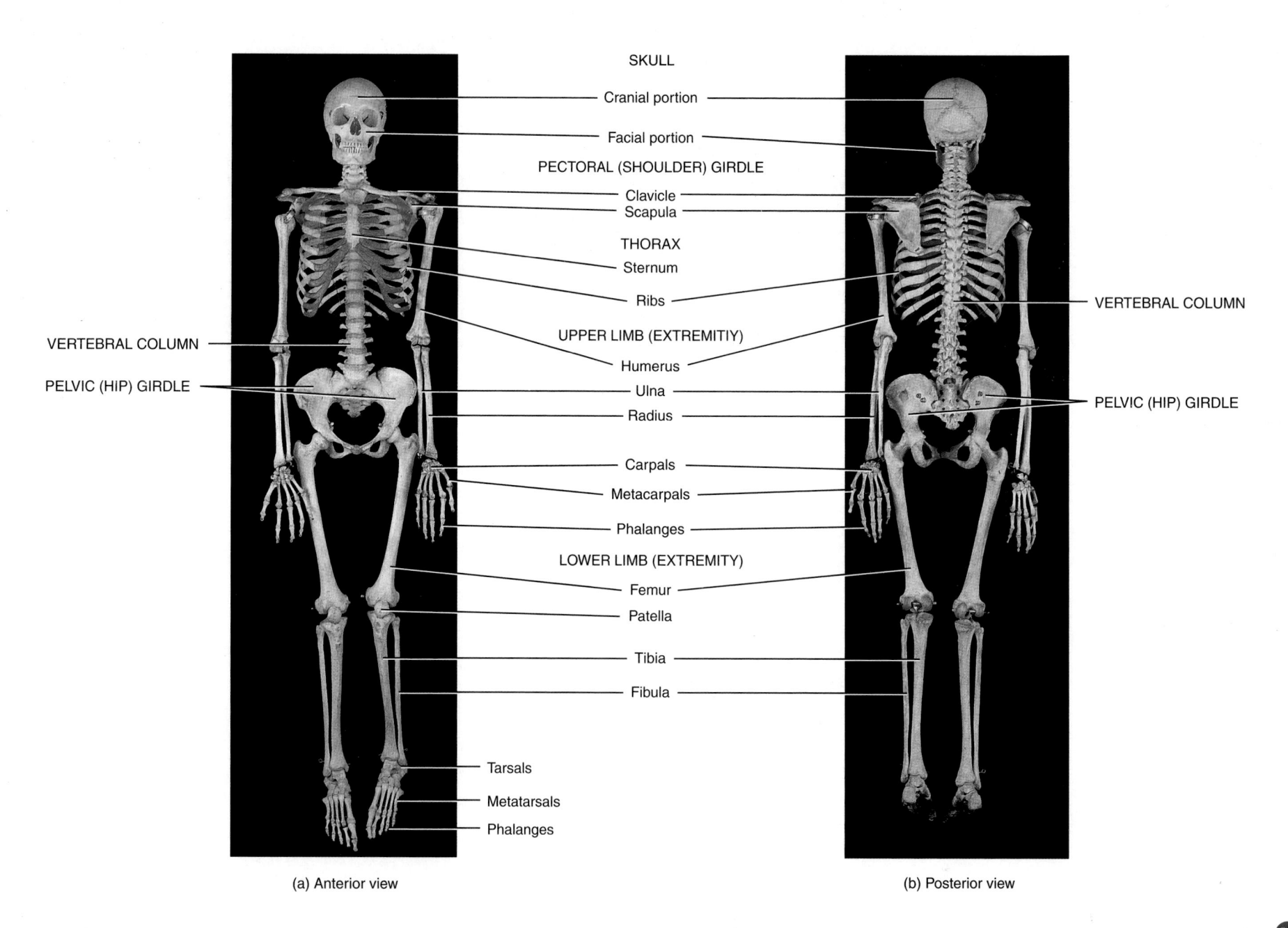

SKULL
Cranial portion
Facial portion
PECTORAL (SHOULDER) GIRDLE
Clavicle
Scapula
THORAX
Sternum
Ribs
UPPER LIMB (EXTREMITIY)
Humerus
Ulna
Radius
Carpals
Metacarpals
Phalanges
LOWER LIMB (EXTREMITY)
Femur
Patella
Tibia
Fibula
VERTEBRAL COLUMN
PELVIC (HIP) GIRDLE
VERTEBRAL COLUMN
PELVIC (HIP) GIRDLE
Tarsals
Metatarsals
Phalanges

(a) Anterior view

(b) Posterior view

FIGURE 35 Complete skeleton